Suzie Two

A Numberline Lane book

by

Fiona and Nick Reynolds

Suzie Two had just returned to her house, having run all the way up Numberline lane and back again.

She started at her own house, Number Two, and counted on as she ran past each house.

"2, 3, 4, 5, 6, 7, 8, 9, 10," she said, before she stopped for a rest outside Jenny Ten's house.

Then she counted backwards as she ran back to her own house again.

"10, 9, 8, 7, 6, 5, 4, 3, 2."

Suzie Two loved running.

She ran everywhere!

Later that morning she ran to see Nigel Nine to buy a packet of counting crisps.

Then she ran to see Clive Five to organise a game of football.

Then she ran to see Hebe Three and had a swim in her swimming pool.

Suzie Two suddenly had an idea.

A lot of the other Numbers never did any exercise.

Suzie Two thought it would be good for them if she organised a Sports Day to help them to get into shape.

It would need lots of careful planning.

It could all take place at Hebe Three's house, as she had plenty of space for all the running, jumping and throwing events.

Suzie Two put up posters along Numberline Lane to tell the others about the Sports Day.

Clive Five was very excited about the idea and he started training straight away.

Even Linus Minus began to practise his jumping skills.

Eventually the big day arrived.

All the Numbers, Linus Minus and Gus Plus arrived at Hebe Three's house.

The first event was the three legged race.

Walter One was running with Nick Six, Nigel Nine was running with Katy Eight and Hebe Three was running with Clive Five.

Gus Plus blew his whistle and all the Numbers rushed off as fast as they could go.

As they came towards the finishing line it was Nigel Nine and Katy Eight who were in first place.

Clive Five and Hebe Three came in second place and Walter One and Nick Six came third.

FINISH

Next came the jumping competition.

Linus Minus, who had been very excited all day, was allowed to go first.

He ran towards the sand pit and managed a super jump from one side to the other.

Then came Clive Five, who was determined to win this event.

He ran with all his might and threw himself across the sand pit, but he didn't quite beat Linus Minus.

Kevin Seven was pleased that he came third.

Clive Five managed second place, but Linus Minus could hardly control his glee as he had come first!

The last event was the egg and spoon race.

All the Numbers and Linus Minus lined up at the starting point ready to go, with their eggs carefully balanced in their spoons.

Gus Plus started the race.

"Ready, steady, go!" called Gus Plus.

Walter One took a quick lead.

Poor Jenny Ten kept dropping her egg.

Linus Minus made a sudden dash and crossed the finish line first, but he had cheated!

He had stuck his egg to his spoon when no one was looking!

By the end of the race, Suzie Two noticed that the strangest thing had happened.

The Numbers had all crossed the finish line in order!

Walter One came first,
Suzie Two came second,
Hebe Three came third,
Nora Four came fourth,
Clive Five came fifth,
Nick Six came sixth,
Kevin Seven came seventh,
Katy Eight came eighth,
Nigel Nine came ninth, and
Jenny Ten came tenth.

All the Numbers agreed that they had really enjoyed the Sports Day.

Nigel Nine handed out glasses of Sum Squash.

Nora Four had brought some sticky cakes.

They all enjoyed a rest at the end of a long and tiring day!